AN EYE FOR FRACTALS

A GRAPHIC & PHOTOGRAPHIC ESSAY BY MICHAEL McGUIRE

ADDISON-WESLEY PUBLISHING COMPANY
The Advanced Book Program
Redwood City, California • Menlo Park, California • Reading, Massachusetts
New York • Don Mills, Ontario • Wokingham, United Kingdom • Amsterdam
Bonn • Sydney • Singapore • Tokyo • Madrid • San Juan

Publisher: *Allan M. Wylde*
Production Manager: *Jan V. Benes*
Promotions Manager: *Laura Likely*
Cover Designer: *Iva Frank*
Production Assistant: *Karl Matsumoto*
Special Advisor: *Robert L. Devaney, Boston University, MA*

This book was typeset by the author.
Camera-ready copy was produced on Linotronic 300.

Cover photo by M. McGuire and cover graphics from an L-system algorithm by P. Prusinkiewicz.

All photographs in this book by M. McGuire except "Mount Williamson, Sierra Nevada"
by Ansel Adams on page 113.

Library of Congress Cataloging-in-Publication Data
McGuire, Michael (Michael D.)
    An eye for fractals: a graphic/photographic essay/Michael McGuire.
        p.cm.
    Includes bibliographical references.
    1. Fractals.   I. Title
QA614.86.M34   1991
514'.74 – dc20   91-47632
ISBN 0-201-55440-2   CIP

ABCDEFGHIJ-KP-943210

To the memory of Mac and Hilda

Foreword

by Benoit B. Mandelbrot,
IBM and Yale University

For years I have been hoping that someone would undertake a work like this. Now the wait has been rewarded by Michael McGuire's photographs. They are a feast for the eye. He is to be congratulated for his initiative, and I recommend his book very warmly.

In addition the author has shown great wisdom in allowing himself only a modicum of well-chosen explanatory words, and there should have been no reason for me not to follow this excellent example. But I did commit myself to write a proper Foreword. Thus allow me to continue by addressing some questions that are bound to jump to the minds of those who open this book:

What is the point of fractal geometry?
What brings together all the diverse pictures in this book?

These two questions are easiest to answer if answered together. With the photographs collected in this book, McGuire proclaims a simple but surprising message. Its sounds a bit grandiose, but this statement remains consistent from source to source. Whether it comes from a sophisticate or from simple folk, it invariably warms my heart.

Fractal geometry is not just a chapter of mathematics,
but one that helps Everyman to see the same old world differently.

Geometry has helped the painter and Everyman see the world in perspective, but this was an exception, and it happened long ago. As a rule helping Everyman and being involved with what he sees are not priority tasks for mathematics as a profession. The resulting inauspicious odds were well known twenty years ago, when computer graphics meant tracing tables, and when the idea that eventually created a need for the word *fractal* had not yet occurred to anyone. This is when I started using the makeshift graphics of that day in the course of research work. The reason for me to rush to graphics was because I am among those who reason best on what can actually be seen. A reason for relying increasingly on graphics was that something strange and totally unexpected began to happen with consistency. Our graphics did more than inform. They made people dream. Colleagues flocked to tell us that we had made them see their own work in a different light, and had helped them by unveiling previously unnoticed analogies. For the first time, they felt that what they saw directly affected what they did next.

We were all astounded by the sudden revelation that the output of a very simple, two-line generating formula does *not* have to be a dry and cold *abstraction.* When the output was what is now called a fractal, no one called it *artificial.* To the contrary, it tended to be viewed as *natural* or (the highest of compliments) as *organic* or as exhibiting some *vital strength.* That is, my work persisted in falling right between the two traditional beliefs, that "simple rules can only generate simple effects" and that "in order to generate complexity, one needs complicated rules." In my youth, everything complicated was reported to require something beyond well travelled physics, a *vital impulse,* to translate poorly the *élan vital* dear to a philosopher named Henri Bergson. Fractals suddenly broadened the realm in which understanding can be based on a plain physical basis.

Do I seem to imply that each of the objects photographed by McGuire can be represented by a two-line formula, or that all are fractal in some near mathematical meaning of this word? Of course not! It continues to be true that there is far more in the world than can be codified within any single geometry. But an additional geometric language is a boon, and its availability helps one appreciate even those forms that lie beyond its rule. All problems are not resolved, but many problems are deeply changed.

Allow me a few more words addressed to the first of the questions raised earlier. *Is there a broader point to fractal geometry?* By itself this question is hard, because it is both the fortune and the misfortune of fractal geometry that it happens to have many distinct points. One might even argue, in jest, that this is an eminently fitting situation, since having many points is very typical of a fractal's shape! These points are of course intimately related, and all are equally worthy in my eyes, but the very fact of having many points does not help universal and unalloyed acceptance. To the contrary, life keeps deepening my awareness of the Old World wisdom of the French Air Force sergeants of my youth, who were wont to proclaim (loudly, of course!) that when a soldier gives several reasons for something, then every one of these reasons must be equally suspect. This wisdom is, of course, widely followed, and it may even be statistically true. But it is the bane of every interdisciplinary effort and of every intellectual synthesis. These sergeants, therefore, have taught me that it is prudent to keep to one reason at a time.

This is why I strongly advise those who open this book to enjoy it solely on its own—sufficiently exalted—terms. Here and now, it does not matter that fractal geometry has also become a nearly routine tool in statistical physics, and is steadily extending its role in other experimental sciences. It does not matter that to many down-to-earth applied scientists the ability to see the world differently has meant having a geometric language at their disposal to make sense of situations where—literally—one could not see anything.

Here and now, it does not matter, either, that fractal geometry has helped trigger a change of mood in parts of mathematics. A resort to illustration was viewed as strictly prohibited only a few years ago, but it has now become routine to many of the best practitioners. The reason is that it has proven a rich source of new mathematical facts, and therefore of new conjectures. It does not matter either that fractal geometry has somehow been granted the hard-to-live-with role of helping to convince Everyman and Everychild (whatever his past level of motivation) that mathematics is a living endeavor. First, it does show that new mathematical facts can be very surprising and that the search for them can be great fun. Second, one soon finds that access to new facts is an indispensable step, but only a first step. It soon becomes imperative to try to organize them. Mathematical proof does not at all reduce to its caricature of making legal and official some facts that had been obvious all along.

Now we come to the arts. Here and now it does not matter that several great composers have seen in fractal structure the features that distinguish music both from collections of indefinitely-held pure notes and from true noise. It does not matter that new people are coming forth who are blessed with equal facility in computers and art, and show promise of creating a new plastic medium, one that will be neither photography nor painting.

At any place and time only one thing matters. Here and now it concerns photography. His *eye for fractals* has enabled Michael McGuire to offer us a truly beautiful volume. Enjoy it.

Preface

This book is about a way to *see*. It is based on a geometry that transcends the points, lines, and planes of Euclid to grasp and describe the shapes of trees and mountains and clouds. Complexity and simplicity are complementary parts of its whole. The geometry is called *fractal* geometry. As formal mathematics, its roots go back about a century when Weierstrass, Cantor, and others discovered some highly irregular curves and sets. These discoveries precipitated something of a crisis in mathematics which was mostly resolved by about sixty years ago. These very irregular objects were then relegated to the mathematicians' nightmare closets as horrific imaginings, certainly of no interest to practical users of mathematics like natural scientists, and not at all a polite thing to bring to the attention of artists.

Then came Benoit Mandelbrot. Knowledgable about this mathematics, he had interests in an eclectic collection of irregular phenomena such as flow records of the river Nile, commodity price fluctuations, and the shapes of coastlines, and made the connection. Over a period of twenty years from about 1960, he built up an understanding that this mathematics, far from being a nightmarish abstraction, was profoundly useful for such matters, and beautiful. He coined the word

from the Latin *fractus,* broken, and three books appeared progressively: *Les Objets Fractals: Forme, Hasard et Dimension* in 1975, *Fractals, Form, Chance and Dimension* in 1977, and *The Fractal Geometry of Nature* in 1982. Mandelbrot's ideas have been taken up enthusiastically in the natural sciences—to the extent that pleas have been heard from journal editors for papers not about fractals!

Two parallel and interrelated developments over the last twenty years have brought fractals to the forefront: the theory of deterministic chaos, and advances in computer graphics. *Strange attractors* and *Julia sets* are visualizations, often quite beautiful, of chaotic behavior which turn out to be fractals. Computer graphics is the indispensable tool for displaying the intensely intricate and iterated structures of fractals.

Mathematicians have known something of fractals for about a century, but artistic consciousness of them is older. We will see that Leonardo had something to say about them, and something of their place in classical Asian landscape painting. Almost any foray into art history turns up examples.

While I was a graduate student in physics at

the University of Washington in Seattle in the early 1970's, I became involved in the climbing and mountaineering way of life that the region offered. This led me into photography as an avocation. The western landscape style of Adams, Weston, and the rest inevitably became my style, both because of its esthetic appeal and because the systematic technical methods that Adams taught made wonderful sense to the scientific side of me. I became proficient in the methods—and then bored. There seemed to be many others equally proficient. Workshops that I attended seemed to focus on finding flaws in images rather than images themselves. Working in color for a while, I found that I was doing light and shadow, the subject matter of black and white.

Mandelbrot's second book was my first contact with fractals. I acquired it in 1977 because of claims that it described a geometry of natural objects. This appealed to me both as a photographer and as a physicist. It made a lot of sense, but nothing much happened until I started doing some programming to generate fractal graphics. As I did this, my eye for fractals in nature cleared and sharpened. I think this happened because to devise an algorithm, especially for a graphical object like a fractal, one must first get a sense of it as a whole, a right-brain perception of it, before plunging into the left-brain process of implementing it in lines of code. Fractal geometry became for me a

rich source of personal inspiration and direction as a photographer. I set myself the goal of writing this book to communicate it. The first attempt choked on verbiage. It is not a subject for a lot of words; it is visual and graphic. Many pieces fell into place when I emphasized that approach. But there are a few spots where some restrained mathematical rigor reveals further beauties of the subject for those who can follow it. I protest the "folk wisdom" that innumeracy is a necessary correlate of artistic ability and talent.

Thanks go to Benoit Mandelbrot for the marvelous and rich concept of fractals and their geometry; to Robert Devaney for especially helpful critical readings of the manuscript; to Heinz-Otto Peitgen for encouraging me to write my original "Eye for Fractals" essay, which appeared in his *Science of Fractal Images;* to Alan Dodds for applying his architect's eye toward improving my graphic presentation; to Julia Siebel and Lynn Simmons for their readings and comments on the manuscript; to my friends at Gallery House Artists' Cooperative in Palo Alto for their long time encouragement of my photography; and to Pat, my wife, who helped me recognize when the time had come to write this book, and who endured me while I did it.

Michael McGuire
October 1990

AN EYE FOR FRACTALS

Consider the images on the next eight pages....

They are different, but how are they related?

Aspens, Wheeler Peak, Nevada

Kelp, Big Sur Coast, California

Pahoehoe #1, Hawaii

Cloud and Chamisa

Buttress, Cathedral Valley, Capitol Reef, Utah

Tree Fern #1, Dandenong Ranges, Australia

Ohia Forest, Na Pali Coast, Kauai 9

Rock Form #1, Pt. Lobos, California

There is...

a kinkiness about aspens,

a swirliness about kelp,

a ropiness about pahoehoe lava,

and a "somethingness" about ferns that wants a common name, a common language. Such a language is *fractal geometry*. Let's see what it is about.

To make a fractal from a triangle,

draw lines connecting the midpoints of the sides,

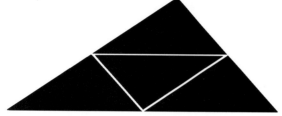

and cut out the center triangle.

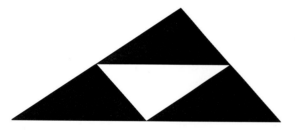

Take the result and do it again,

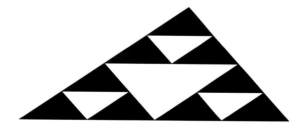

and again,

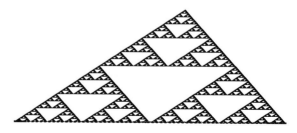

and again,

and again,

forever iterate.

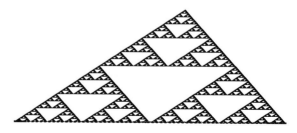

the Sierpinski triangle

A fractal

looks the same

over all ranges

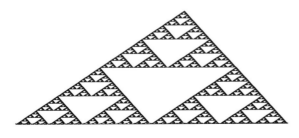

of scale.

This is called "self-similarity."

Self-similarity

over limited ranges of scale

is common in nature.

Some words used with fractals—

Iterate: to repeat an operation, generally using the last result of that operation as the input.

Self-similarity: as in the example, a similar appearance at all scales. The word similar later will need a more general understanding than the strict euclidian sense that is true in the example.

Fractal dimension: Fractals pack an infinity into "a grain of sand." This infinity appears when one tries to measure them. The resolution lies in regarding them as falling between dimensions. The dimension of a fractal in general is not a whole number, not an integer. So a *fractal curve,* a nominally one-dimensional object in a plane which has two dimensions, has a fractal dimension that lies between 1 and 2. Likewise, a *fractal surface* has a dimension between 2 and 3. The value depends on how the fractal is constructed. The closer the dimension of a fractal is to its possible upper limit which is the dimension of the space in which it is embedded, the rougher, the more filling of that space it is. More on this after some more examples of fractals.

Replacement rule: In going from one stage of construction of a fractal to the next, one graphical object is replaced with another, which is usually more complex, but which fits into the place of the original. For example, replace a straight line segment

with this figure of four line segments each of which is 1/3 the length of the original.

Now apply that rule to this triangle.

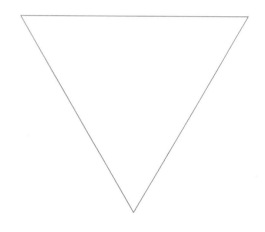

14

to get this result:

After another iteration:

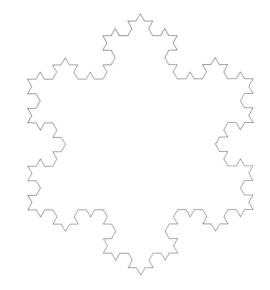

Apply the replacement rule again to the above result (iterate) to get

After infinitely more iterations we get the von Koch snowflake.

In this replacement rule, the figure below is both the starting figure and the shape that is replaced.

Below are the second and third iterations.

Below is the replacement figure, which is in fact the first iteration. In further iterations, each instance of the starting shape is replaced by an appropriately scaled, turned, and if necessary, flipped-over copy of this replacement rule figure.

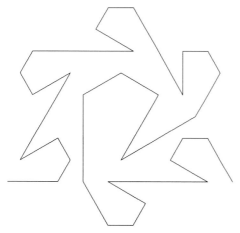

After another iteration (and enlarging), we get this figure by Mandelbrot, inspired by Peano and von Koch.

Peculiar things happen when one tries to measure fractals. The von Koch replacement rule for every iteration increases the length by 4/3.

After infinite iteration, the distance between the endpoints along this von Koch "arc" would be infinite.

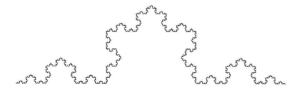

Keeping in mind that this

is really this,

try doubling the dimensions of this Sierpinski triangle,

to get

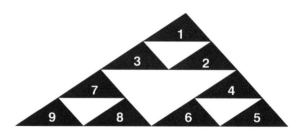

We seem to have increased the amount of surface, the area, by a factor of 3. This is odd. Doubling the dimensions of a non-fractal plane figure such as a square or a circle or a triangle increases its area by 4 times. Does the Sierpinski triangle have any area at all? In the infinite iteration limit its area vanishes. Yet it is certainly occupying two-dimensional space. So it is plausible that it is somehow between one and two dimensions. To clarify this, consider the dimensional behavior of some things we understand: a line, a square, a cube. Let's divide them into pieces whose linear dimension is $1/n$ the size of the original. So there are n pieces of the line,

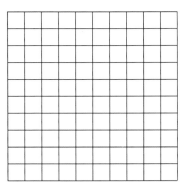

and n^2 pieces of the square,

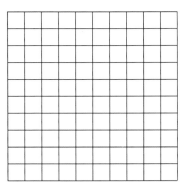

and n^3 pieces of the cube whose sides are length $1/n$.

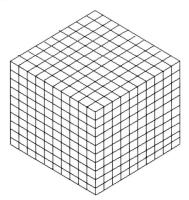

These pieces are of course trivially self-similar. Notice that the exponents of n, 1 (which as a matter of notation is usually not explicitly written), 2, and 3, are the dimensionality of the objects to which they refer. Now readers who have thankfully forgotten about logarithms may pass over the rest of this page and the next. For the rest, here is a reminder of the relation between logarithms and exponents:

$$\log(a^b) = b \log(a).$$

So for the line of n pieces

$$\begin{aligned} \log(\text{number of pieces}) &= \log(n^1) \\ &= 1 \log(n), \end{aligned}$$

and for the square with a side of n pieces,

$$\begin{aligned} \log(\text{number of pieces}) &= \log(n^2) \\ &= 2 \log(n), \end{aligned}$$

and for the cube with an edge of n pieces,

$$\begin{aligned} \log(\text{number of pieces}) &= \log(n^3) \\ &= 3 \log(n). \end{aligned}$$

Now since the pieces are self-similar, a magnification of a piece by n will give us back the object. So the dimension D is given by

$$D = \log(\text{number of pieces}) / \log(\text{magnification})$$

Thus for the line,

$$D = \log(n^1) / \log(n)$$

$$= 1 \log(n) / \log(n)$$

$$= 1,$$

and for the square

$$D = \log(n^2) / \log(n)$$

$$= 2 \log(n) / \log(n)$$

$$= 2,$$

and for the cube

$$D = \log(n^3) / \log(n)$$

$$= 3 \log(n) / \log(n)$$

$$= 3.$$

Recall now for the Sierpinski triangle that when we magnified by 2 we got a threefold increase in the number of triangles. So

$$D = \log(3) / \log(2) = 1.58...$$

In the Koch snowflake a step of iteration is to magnify by 3 and get 4 pieces, so its dimension is

$$D = \log(4) / \log(3) = 1.26...$$

The Mandelbrot-Peano-von Koch snowflake

is a case where D approaches a limiting value of 2.0. Compare the three example fractals and notice that the greater D is, the more the fractal fills the space within its boundary.

The question of two pages ago—how "long" is the von Koch arc—now has another answer. Recall that its length (one-dimensional measure) is infinite. But as a set of line segments it occupies zero area (two-dimensional measure). It turns out that the dimension in which it is neither zero nor infinite is this strange non-integer dimension of 1.26... To see this, take L_0 as the straight-line distance between the endpoints. If the stage of the constructions is such that it consists of n pieces of length d, then the magnification is L_0 / d and

$$D = \log(n) / \log(L_0 / d).$$

This can be rearranged as

$$L_0^D = nd^D.$$

On the left is the distance between the endpoints raised to the power of the fractal dimension. As we pass to the limit of infinite iteration where n becomes infinite and d becomes infinitesimal, we get this measure of the fractal in its dimension which is neither zero nor infinite.

There are fractals that fall between 0 and 1

in their dimension. One such fractal is the Cantor middle-third set. It is constructed by iteratively cutting out the middle thirds of line segments, a one-dimensional analog of the Sierpinski triangle.

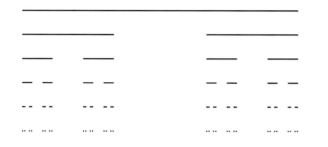

It ends up as a dust of points, curdled about certain locations. Since a magnification by 3 gets 2 pieces,

$$D = \log(2) / \log(3) = 0.63...$$

A cross section perpendicular to the plane of the rings of Saturn seems to have this style of fractal structure.

Fractals are visually complex mathematical objects that are similar in shape and structure over an *infinite* range of scale. They arise from the iterated application of a simple rule. Nature is visually complex and has material shapes and structures that are similar over a *finite* range of scale. Over this finite range of scale with self-similar structure, we might analogously expect a simplicity in the rule or law of formation or growth that produced the structure. This could be a constancy of the underlying geology of a mountain, or the genetic identity of the DNA molecule that programs the growth of a tree. Erosion and growth are in some sense iterative.

Adding a little randomness to the Sierpinski triangle can produce something very natural looking. Let's change the rule so that instead of using the exact midpoints of the sides of the triangle, we take a point at random around the midpoint. The self-similarity will be statistical rather than absolute. Let's constrain the randomness by requiring the point to be within some distance from the midpoint; make it somewhere within a circle centered on the midpoint whose diameter is half the length of the side. Connect these points and the corners. The heavy lines are the starting triangle.

The heavy lines are the result of the last iteration.

Clear the construction and make the corners black.

After eight more iterations, and enlarging:

Cliff, Salt Point, Mendocino Coast, California

Shiprock, New Mexico

24

Pahoehoe #2, Hawaii 25

Cliff, Wheeler Peak, Nevada

26

Zabriskie Pt., Death Valley, California

Marion Lake, Sierra Nevada, California

There is more to say about fractal dimension. A fractal figure lying in the plane has a dimension between 1 and 2. The closer it comes to the upper limit of 2, the more "spacefilling" it is. The notion extends readily to surfaces in three-dimensional space. A piece of aluminum foil fresh off the roll has a fractal dimension of slightly more than 2.0. After crumpling and uncrumpling it has increased toward, but is less than, 3.0. The next three pictures are of 3-dimensional "fractal mountains" generated using the same sequence of random numbers, but varying a constraint to produce a different value of fractal dimension for each picture.

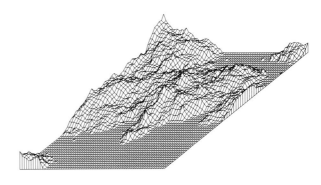

Fractal dimension 2.2

The first of these pictures with dimension 2.9 is impossibly jagged, a caricature of mountains. The second with dimension 2.5 is acceptable artistic exaggeration, while the third with dimension 2.2 is about what is observed in nature. The constraint that produced the differences in fractal dimension was to vary the amount of fine, high-frequency, rapidly changing detail like this

relative to the coarse, low-frequency, slowly changing detail like this.

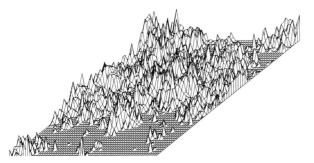

Fractal dimension 2.9

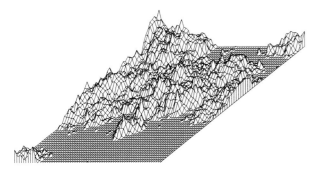

Fractal dimension 2.5

The spectrum or spread in spatial frequency of a landscape is closely related to its fractal dimension. Below are three curves showing

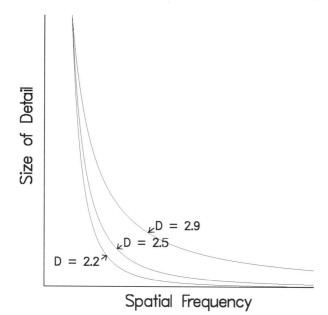

how size of detail varies with spatial frequency for the three values of fractal dimension. Notice how the spectral curve for landscape of dimension 2.9 dies away much more slowly than the others as the frequency increases. That is, the high-frequency jaggedness is greatest.

There are some intriguing implications. One is that the spatial-frequency spectrum of a reconnaissance photograph might reveal something, otherwise well camouflaged, by a departure from the expected fractal characteristics. Another relates to a concept among perceptual psychologists that sen-

sory processing is "efficient" in the way it encodes and handles information presented to it. And fractals are nothing if not efficient in generating visual complexity from the least amount of information. So a brain shaped by evolution in a natural world might be expected to have structures complementary to, and adept at apprehending fractal structure in nature. Some recent work points in just this direction. On the right is a fractal structure based on hexagons which has been proposed as a possible way the receptive fields of the brain's visual cortex are organized. The smallest hexagons correspond to the close-packed structure of the cells of the retina and the perception of finest details. The layers of larger hexagons organize the underlying layers to respond to progressively coarser detail. And further, this organization seems to fit the spectra of natural images.

The mathematicians who first came across fractal objects about a hundred years ago considered them monsters of uncompromising irregularity. Mandelbrot, remarking on this and the fractal characteristics of arterial networks and the air channels of the lungs, commented that "Lebesgue-Osgood monsters are the very substance of our flesh!" Now it seems that they may have much to do with the way we see and think.

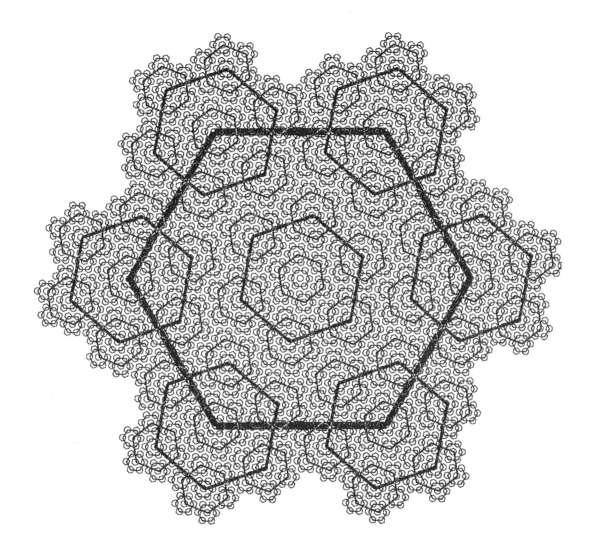

31

Afternoon, North Cascades, Washington

Downey Peak, North Cascades, Washington   33

Mt. Clarence King, Sierra Nevada, California

From Gunsight Pass, North Cascades, Washington

Sulfur Creek, Capitol Reef, Utah

Needles, Canyonlands, Utah

But randomness is not required. A tree can start with a simple branching rule. From a trunk grow two branches. The lengths of the branches are to be less than the length of the trunk, and the angles between the line of the trunk and the branches are given. Here we have a branch at 10°, 9/10 the length of the trunk and one at 60°, 6/10 the length of the trunk.

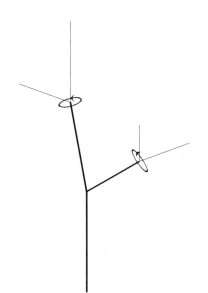

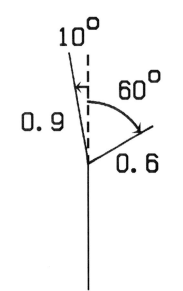

After five iterations...

Now treat the branches as trunks and do the same branching again at their ends. If at the first iteration we rotate the planes of these new branchings by 90°, the figure becomes three-dimensional.

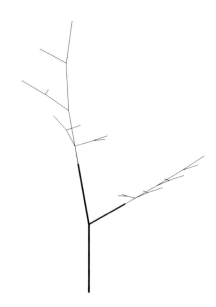

After ten iterations, view from the front.   After ten iterations, view from the top.

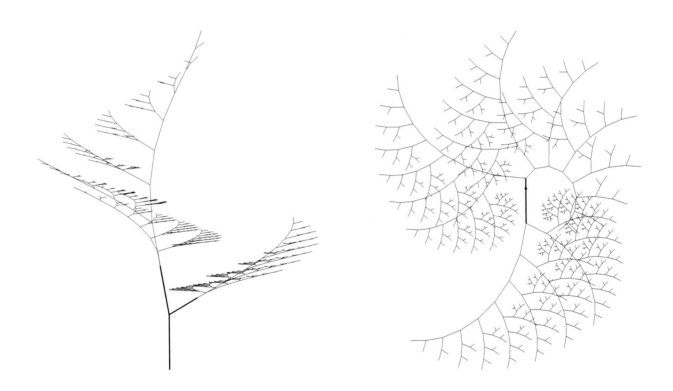

39

Oak #1, Arastradero Preserve, Palo Alto, California

Trees, Merced River, Yosemite, California   41

Elephant Tree, Anza Borrego Desert, California

Trees and Cows, Carson Valley, California

43

Generate a bush fractal with this replacement rule.

The second and third iterations...

After two more iterations, and enlarging...

Smoketree, Anza Borrego Desert, California

Chamisa near Lee Vining, California

47

Kelp, Pt. Lobos, California

48

Roots, Sixty Lakes Basin, Sierra Nevada, California

50   Trees at Happy Isles, Yosemite, California

The graphic fractal examples shown so far were all based on *line replacement* rules. A less obvious but very rich approach to fractals is based on *transformation replacement*. A transformation as it is meant here relates the position of a point in an image or on a map to another position in the same or different image or map. The lens of a camera transforms or maps a point in the landscape to its corresponding point on the film. The transformations here are much simpler.

There are two major kinds of transformations of fractal interest, linear and nonlinear. Both can produce fractals, but of rather different types. To demonstrate linear transformation, let's start with an image, an uppercase "R" drawn in a square. A transformation consistently changes all the points in the image in the same way. What the transformation does to the shape of the square is what it does to all the points in the image.

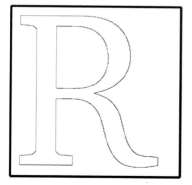

Some linear transformations to this image are to make it smaller

or larger,

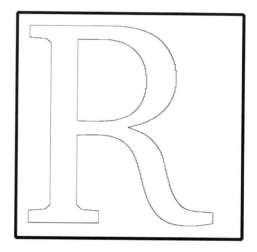

scale it horizontally or vertically,

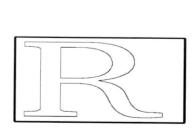 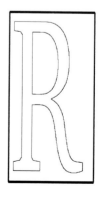

scale it and rotate it or shear it,

or invert it by reflection.

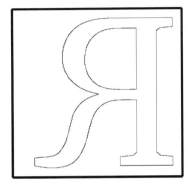

These particular linear transformations, which can do any combination of moving, rotating, scaling, shearing, or inverting as long as the opposite sides of a square it would transform end up parallel, are called *affine*. Neither the squares nor the R's are the transformations themselves but merely show the effects of them. The results are no longer quite self-similar; they are self-affine, a more general relationship.

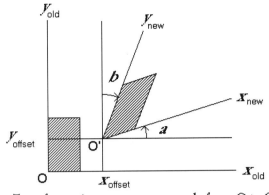

Transformation moves a rectangle from O to O', rotates it, and shears it to a parallelogram

While not necessary to understand what follows, the mathematical form of these affine transformations in two dimensions is

$$x_{new} = r \cos(a) \, x_{old} - s \sin(b) \, y_{old} + x_{offset},$$

$$y_{new} = r \sin(a) \, x_{old} - s \cos(b) \, y_{old} + y_{offset}.$$

The scaling factors along the x and y axes are r and s, and the angles between the old and new x and y axes are a and b respectively. Two special cases are where $a = b$ and $r = s$, the similarity transformation (results are self-similar), and where $a = b$ and $r = s = 1$, which is the rotation transformation. The process of transforming an image is to impose a grid on it, then compute and mark the new location of each pixel.

Now let's apply several contracting or down-sizing transformations at a time and iterate. Below are three transformations of the original R-square which half-size it, and position themselves with respect to each other as shown.

And again using the image from the last iteration, apply the three transformations again to get

Now using the three transformed images together as the original, apply the three transformations again to get

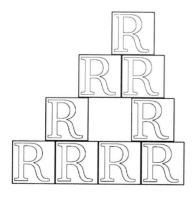

Again I emphasize that the R's and the squares are not the transformations and have nothing intrinsic to do with the Sierpinksi triangle that has emerged from the iteration. They show the effects of the transformations.

Any starting shape—a circle, an asterisk, a single point—on which these transformations were iterated would produce this result. This subtle idea was discovered by John Hutchinson, a mathematician at the University of Canberra in Australia, and was applied in interesting ways by Michael Barnsley and his associates at Georgia Tech. But there are easier ways to get Sierpinski triangles. This approach is interesting for what *else* it can do.

Next are four transformations, again demonstrated by their effects on the R-square. The least obvious of them shrinks the R-square to the short horizontal line at the lower left.

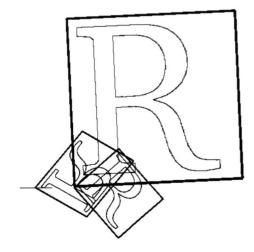

Now apply the four transformations to the last image to get

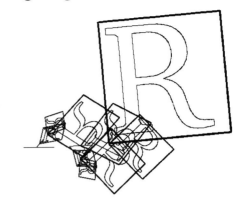

After four more iterations, something is emerging...

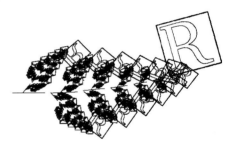

Sufficient iteration will shrink the R-square onto the final shape, the black spleenwort fern, which has become Barnsley's trademark:

This shape is the *attractor* of iteration of this set of transformations.

There is a random iteration method for rendering images that is more efficient than direct transformation replacement. Start anywhere on or near the image. Choose one of the transformations at random. Transform the point you are on to a new point and mark it, and choose again. Shortly you will be on the attractor, and will stay on it as you transform. The marks you make become the image. The "dice" for the random choice are loaded proportional to areas of the transformations relative to the total area. Thus the stem transformation won't be chosen as often as the others. Seen on a computer display the process looks like a photograph developing.

But how to find a set of transformations to produce a fractal image? Knowing what will happen, one can perhaps see something of the fern in the first iteration of the transformations applied to the R-square. This is the key. Barnsley has devised an answer in the form of his "collage" theorem. Fractal images are a collage of transformed copies of the whole. Start with an outline copy of the whole as on the left below. Use transformed copies of this outline on the outline itself to more or less cover it like a collage, as those on the right below would begin to do. Use the transformations that will do this to generate the fractal.

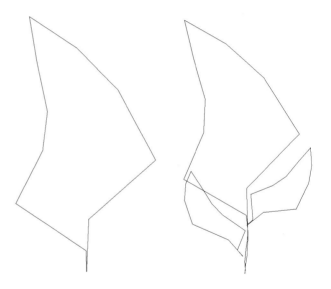

On the next page is the fern that results from the iteration of these transformations.

Tree Fern #2, Dandenong Ranges, Australia   57

Bracken Fern, Sierra Nevada, California

The old-fashioned way to store an image digitally is to impose a grid on it and record the average gray level of each grid cell (pixel). The finer the detail needed, the finer the grid must be and the larger the record.

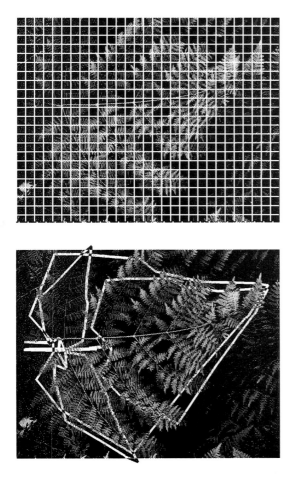

Barnsley proposes a fractal approach based on the collage theorem. Fineness of detail is a matter of persistence in iteration. Above, sketched on the photograph, is an outline of

the fern and collaged upon it are four transformed copies of it. This is the result of iterating the transformations.

For storage purposes, a transformation consists of a list of six numbers. Four transformations—twenty-four numbers plus the random iteration algorithm—is sufficient to generate this interesting start on reconstructing the photograph of the fern. A tenfold or hundredfold increase in the number of transformations would still be minuscule compared to the million or so numbers required for conventional storage. Getting the transformations right is the tricky part. One way to do it is in an interactive computer graphics environment, controlled by a mouse or similar pointing device.

Intelligence can be applied to conventional storage to reduce the space required. A simple thing to do if there is a run of pixels

59

with the same value is to encode them with two numbers—a count of the pixels in the run, and the value common to them all—rather than recording all the values in the run. And there are much more elaborate methods. What they all have in common is that they somehow find and remove redundancies in the image. Fractals are in a way the ultimate in redundancy, so this method is potentially the most effective compression of all. Of course, ferns are more obviously fractal than faces, but it has been done with faces.

The images on the following pages I created using a "transformation editor" program that I wrote, which is based on the collage theorem. My original purpose was to explore this matter of image compression. It turned out to be difficult. Just what to do next after the first four transformations in reconstructing the photograph of the fern is not obvious. Barnsley has founded a company that supplies a proprietary answer to "what to do next." However, the esthetic appeal of what *was* possible to do with it was interesting in itself.

The way it works is that after sketching an outline on the screen with the mouse, I can interactively create, move, rotate, shear, and scale copies of the outline to create a collage of them upon the outline and iterate to see the effect. Outlines of archetypically natural objects like leaves and branchings work well.

After some cycles of interaction and iteration, I get a result which I save to print later on a laser printer at higher resolution. The result could just as well be printed at poster or even billboard size right down to the highest resolution that the printer can address. Unlike a photograph, it will not lose resolution as it is enlarged.

Computer graphics images are mostly produced by techniques metaphorically taken from conventional art such as painting or drawing. The programs used typically have "paint" or "draw" as part of their names. Beyond this there is a kind of synthetic photography which computes an image given an object like the fractal mountains, a light source, and an observer position. But these transformation graphics are rather different. They seem more natively, more naively, the result of the computer itself as an artist's tool without much intervening metaphor.

What incidentally is the creation that might be copyrighted here? Certainly the realized image on the page could be copyrighted, but the real intangible creation is the set of transformations. There is an analogy to a photographic negative, but the question is open.

Included with the images is a postage stamp view of the outline and collage that produces them.

60

61

The next three images use the same starting outline, but different collages of transformations of it.

63

66

Strangler Fig #1, Queensland, Australia

Bamboo, Road to Hana, Maui

Maidenhair Fern, Zion, Utah   69

Uluhe Fern, Maui

Sea Laburnum, Queensland, Australia

Antelope Canyon #1, Arizona

A whole new world of fractals which encompasses the dynamic behavior known as *deterministic chaos* arises from the iteration of simple *nonlinear* rules. Here are some words and ideas to start.

Nonlinear: In mathematical terms take the simple equation

$$y = x^n.$$

If n has any value except one, the equation is nonlinear. Of course if $n = 1$, as a matter of notation, it is not written. Geometrically, if $n = 1$, the equation plots as a straight line, otherwise as a curve, hence the word nonlinear. In terms of human experience, effects are no longer proportional to causes when things go nonlinear. Since $n = 1$ is a single possibility out of an infinity, referring to the study of nonlinear dynamics is a bit like calling zoology, except for elephants, the study of non-elephant animals. Linear studies are popular because the solutions are much easier to obtain.

Real number: Geometrically a real number is any position on a line which extends infinitely.

The line of real numbers

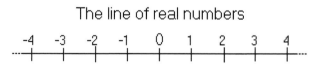

Imaginary number: An imaginary number is the square root of a negative real number.

"Imaginary" comes from imagining that -1 or any other negative number has a square root, a mathematical supposition that is extremely useful. The unit imaginary number is $i = \sqrt{-1}$. By definition, $i^2 = -1$. In a similar geometric sense, there is a line of imaginary numbers.

Complex number: If one combines the real line and the imaginary line at a right angle, a plane is created, the *complex* plane.

The Complex Plane

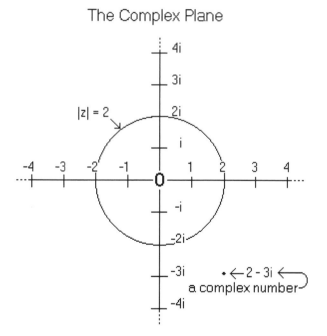

Just as a position on the real line is a real number and a position on the imaginary line is an imaginary number, a position in the complex plane is a complex number. It is written as the sum of a real number and an imaginary number, but the actual addition is not performed. A preferred notation uses x as a variable along the real axis, y as a variable along the imaginary axis, and z as the complex variable, that is, the variable position in the plane so that

$$z = x + iy.$$

A simple transformation of z that is at the heart of nonlinear fractal geometry is to square it and add a constant value to it.

$$z_{new} = z^2_{old} + c.$$

The c is some constant position in the complex plane which consists of say, real part a, and imaginary part b. So writing it all out,

$$z = (x + iy)^2 + a + ib$$

$$= (x + iy)(x + iy) + a + ib$$

$$= x^2 + i^2y^2 + a + i(2xy + b)$$

$$= x^2 - y^2 + a + i(2xy + b)$$

Thus i functions as a kind of bookkeeping device for keeping separate the real and imaginary parts of a complex number, and plays its part in defining the operation of squaring a complex number. A final thing we need is the absolute value of z, a real number which is denoted $|z|$ and has the meaning of the radial distance from the origin or center of the complex plane to z regardless of what direction it is. Thus $|z|$ = 2 can be any point on circle of radius 2 centered on the origin. In fact, it is a definition of that circle.

Attractor: The figures that arise from iterating linear transformations are said to be the attractors of the iteration. A really simple example of attraction is what happens when a number on the real line is iteratively squared; that is

$$x_{new} = x^2_{old}$$

Try starting with $x_{old} = 0.9$

$$0.9 \times 0.9 = 0.81$$
$$0.81 \times 0.81 = 0.6561$$
$$0.6561 \times 0.6561 = 0..43046...$$
$$... \text{ after ten iterations}$$
$$= 1.39... \times 10^{-47}$$

The attractor point of this iteration is zero. Now try $x_{old} = 1.1$

$$1.1 \times 1.1 = 1.21$$
$$1.21 \times 1.21 = 1.4641$$
$$1.4641 \times 1.4641 = 2.14358...$$
$$...\text{after ten iterations}$$
$$= 2.43... \times 10^{42}$$

The attractor point of this iteration is infinity. If we try 1.0, it might appear that the attractor point is 1.0 because the result remains fixed at 1.0. In fact, 1.0 is unstable. The slightest bit above 1.0 and the iteration is attracted to infinity; below 1.0 it is attracted to zero. Try this on a pocket calculator to get a feel for the process.

In the complex plane, $z_{new} = z^2_{old}$ behaves similarly under iteration. If the starting point of iteration is inside the circle, $|z| = 1$, the attractor is zero; if it is outside, the attractor is infinity. If it is on the circle, it hops from place to place on the circle, and is unstable against the smallest change from $|z| = 1$. This is not very interesting. But add a complex constant c to get the equation mentioned above,

$$z_{new} = z^2_{old} + c,$$

and it is hard to find sufficient superlatives to describe the geometric complexity and beauty of the result for some values of c. The shape of the figure which is the boundary between attractor points is no longer a circle (except for the trivial case of $c = 0$), but a fractal. And that fractal shape depends sensitively on c. These boundaries are known as Julia sets after Gaston Julia, a French mathematician who, with his contemporary Pierre Fatou, was the first to study iteration in the complex plane about the time of the first world war.

Following are some Julia sets for different values of c. Julia sets all fall within the circle

$|z| = 2$ and are symmetric about the origin. Points are colored black if, starting at that value of z, the iteration is *not* attracted to infinity. Where there are solid regions of black, the Julia sets are just the boundary.

$c = -0.122 + 0.745i$—Douady's rabbit

$c = -0.754 + 0.049i$

Except for the attractor at infinity, the number and location of the attractors depends on c. In some cases there are no attractors except infinity and no "inside" to the set in the sense that there *is* an inside to the circle of the $c = 0$ case. Then the set is a "Fatou dust," a fractal dust like the Cantor set described earlier, but more interestingly arranged:

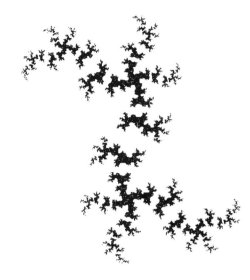

$c = 0.354 + 0.536i$

If our starting point for iteration is one of the points of the set, the iteration will hop around locally from point to point in the set. But if it doesn't start on a point of the set, it will sooner or later run away to infinity. Moving along a continuum of values of c, there will be a place where a dust coalesces to a filament, and then going further begins to have an inside. When it has an inside, there are attractor points other than infinity. The set at right above is just past the filamentary stage and begining to have an inside.

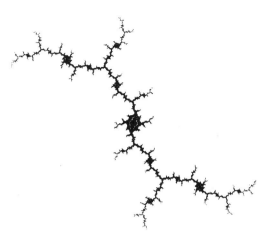

$c = -0.420 + 0.986i$

Following are four sets which pass from the "dust" condition to the solid condition by small changes in the real part of c.

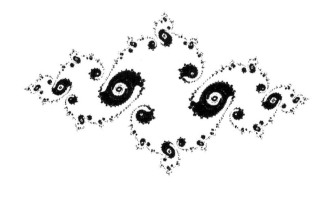

$c = -0.756 + 0.097i$

76

$c = -0.744 + 0.097i$

$c = -0.743 + 0.097i$

$c = -0.736 + 0.097i$

The way a Julia set is mapped out is to use a grid of points in the square region bounded by -2 and +2 on the real axis and -2i and +2i on the imaginary axis. Iterate z for each grid point up to some maximum number of iterations—generally the more iterations, the better defined the result. If the absolute value of z becomes greater than 2 in the process of iterating, then there is a proof that the initial value of z is one that is attracted to infinity, and iteration should cease. If this doesn't happen before the maximum number of iterations is reached, then the point is assumed to be in or within the set and colored say, black, otherwise white. A refinement of this is to keep the count of the number of iterations for each point and use this to create a contour map whose "elevations" are the number of iterations.

$c = 0.354 + 0.536i$

A further refinement is to color this map according to the elevation. This is especially useful for sets that are dusts because they tend to be too diffuse to display otherwise. The dust sets on the preceeding pages have solid black regions because they were under-iterated to avoid their disappearing. Mandelbrot started the work of investigating Julia sets with this computer graphic approach. Important early workers were Adrien Douady and John Hubbard. The premier picture book is Peitgen and Richter's *Beauty of Fractals*.

For what values of c do Julia sets have connected substance? Mandelbrot posed this question. The answer he found was an incredible organizing principle. The *Mandelbrot set*, the object at the right, is the part of the complex plane where the c values have Julia sets which are not dusts.

On the following pages is a zoom into the Mandelbrot set. The explanatory text resumes after it.

$$c = -0.766227 + 0.096990i$$

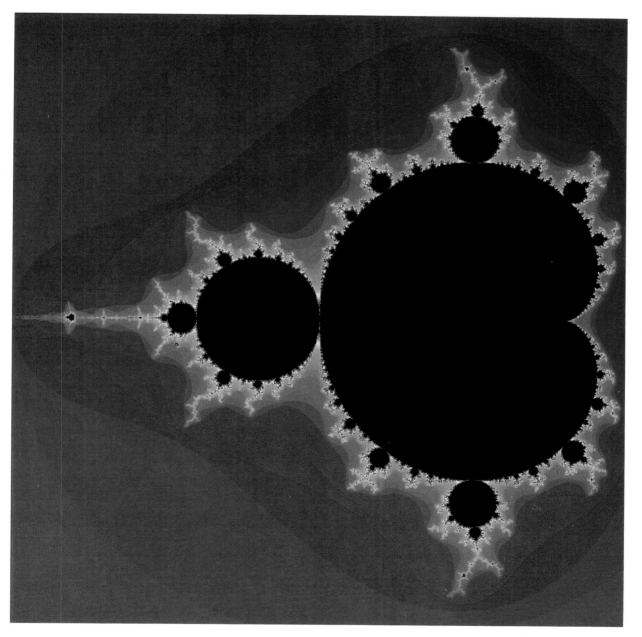

The Mandelbrot Set

79

The Seahorse Valley

The Kelp Pool

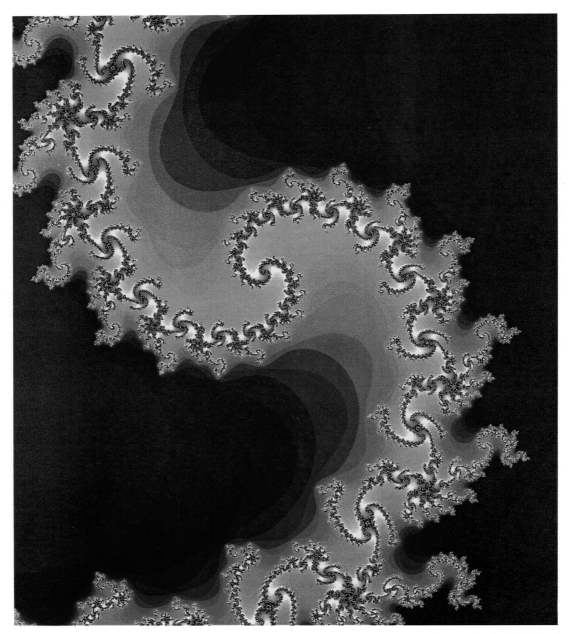

Deeper in the Kelp Pool

Still Deeper in the Kelp Pool

Mandelbrot Miniature from the Kelp Pool

84

Julia Set from the Center of the Kelp Pool Zoom

The Mandelbrot set has been called one of the most complex objects in all of mathematics. It entered popular culture through A. K. Dewdney's article in the August 1985 *Scientific American*. Computing it became a recreation verging upon obsession in world of personal computers. It can be mapped out by techniques similar to Julia sets. One iterates the same equation as the Julia set, but the starting point is always $z = 0$. The test is whether iteration for a given value of c is attracted to infinity or not. There is a similar condition on the absolute value of z becoming greater than 2. Colored contour maps of the iteration counts produce marvelous pictures, perhaps like tidepools on an alien planet. The colored version whole set was on p. 79. This is it shown in outline.

gone after a few iterations. The star-shaped figure is trajectory of a "c" at point b which is just inside the set and never gets away. The small boxed region is what is enlarged in the next picture of the zoom sequence as on p. 80 and below, and the rest similarly.

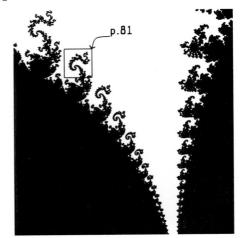

from p. 80

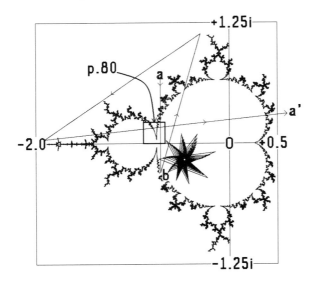

The trajectory aa' is the path under iteration of a "c" at point a from outside the set. It's

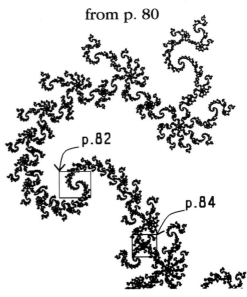

from p. 81

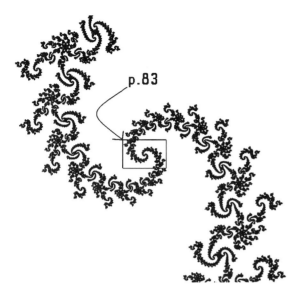

p.83

from p. 82

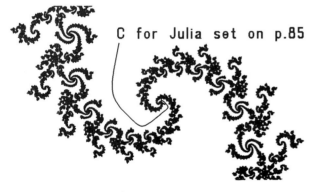

C for Julia set on p.85

from p. 83

The Mandelbrot set is obviously not self-similar on the large scale, but on the small scale it is, and in interesting ways, as the zoom sequence shows. There are miniature copies of itself in the space surrounding the main body, connected to it by vanishingly thin filaments. This is the one that is boxed in the second stage of this guide to the zoom sequence and shown in color on p. 84.

The set is generally most interesting near its boundary, and there is a decided similarity between what one sees zooming in on a region and the Julia sets whose c's are in that region. The final figure in the color plates is the Julia set whose c is centered at the end of the zoom sequence.

Lichen, Big Sur, California

Kukui Grove, Road to Hana, Maui

89

Rock Form #2, Pt. Lobos, California

Canyonlands from Island-in-the-Sky, Utah

Some abstract but playful numerology produces these fractals using nonlinear transformations. Do they have a connection to the natural world as we see it? Not directly, although there is about them in Peitgen and Richter's words "an unending filigreed entanglement, an unceasing bargaining for even the smallest areas" and "diversity that flourishes particularly well at boundaries." The previous four photographs have this about them.

These fractals are a way of visualizing *chaotic* behavior. The word "chaos" is used here in its *courant* sense, that is deterministic chaos. This is in distinction from the older meaning, which implied incomprehensible disorder. It is a behavior that is in some way fractal. The complexity of the behavior arises from the iteration of simple rules. While it need not create a visible fractal structure, it can leave tracks in the natural world which are fractal, for example turbulent flows and their erosion patterns.

There are a variety of ways that chaos can arise, but about the simplest is the squaring process of the Julia and Mandelbrot sets. It appears in iterative growth processes such as the compounding of interest or the growth of biological populations. At first glance such a process seems to be linear:

$$x_{new} = (1 + R)\, x^1_{old}$$

The exponent 1 is an explicit reminder that it is linear. The new value of the account or the population, x_{new}, is linearly related to the previous value, x_{old}, by an apparently constant interest or growth rate R. The result of iterating this process is the well-known exponential growth, which may be biologically or financially catastrophic but is not chaotic. The nonlinearity comes in because exponential growth in real life eventually reaches a limit, and the rate can no longer be constant. but is affected by size of the population or the account. A biological example is a population of fish in a pond with a fixed annual food supply. If the population exceeds what the pond can sustain, some or all of the fish will die. The growth rate itself then depends on the population. Financially there is the old joke that owing the bank $50,000 and not being able to pay is *your* problem; owing $50 billion is the *bank's* problem. In the latter case what you owe is of the order of the size of the bank's assets, and it affects the interest rate the bank can pay or can expect you to pay.

One expects that there is some maximum population in the pond that can be sustained, M. The variable growth rate R might depend on this and the actual population x in the following way

$$R = r\,(M - x).$$

If x is less than M the growth rate is positive

and population increases; if greater than M the growth rate is negative and the population decreases. We can scale the populations in terms of M, setting M to 1. Then we take r as the constant growth rate if the food supply were infinite or, equivalently, the population vanishingly small. Then the process looks like

$$x_{new} = [1 + r(1 - x_{old})] \, x_{old}$$

$$= (1 + r) \, x_{old} - r \, x^2_{old}.$$

This is sometimes called the logistic equation. The exponent 2 appears, telling us that this is nonlinear. Iterating this simple growth law produces a variety of behaviors, depending on the value of r. There are two special case starting points that go nowhere, $x = 0$ and $x = 1$. They are unstable in that the slightest difference from these values does go somewhere. Up to $r = 2$, what happens is not very surprising. The population, x, is attracted to M, which we scaled to 1, regardless of its starting point aside from the special cases. It performs a damped oscillation above and below as it settles onto the attractor over fifty iterations.

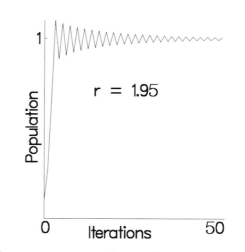

Then some surprises—for r greater than 2 the population cycles periodically between two values. The attractor is not a single point, but this cycle that goes between two values, again independent of the starting point. Below we start at $x = 0.1$.

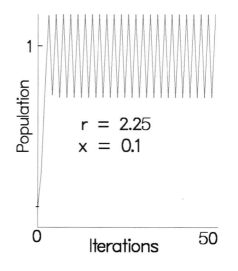

And here we start at $x = 0.5$, same r as

before. The same cycle attracts, even though the start values are well separated.

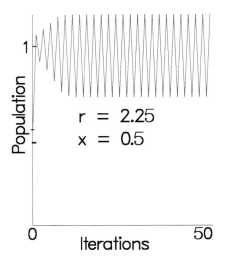

And for r greater than $\sqrt{6} = 2.449...$ it cycles periodically through four values,

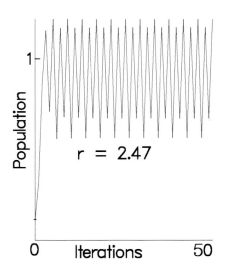

and then eight, sixteen, thirty-two.... This is called a *period doubling cascade* because with each step in the number of cycles, the number of iterations required to return to a given value in the cycle is doubled. Finally for r greater than 2.570 the periodicity disappears, that is the number of cycles becomes infinite, which is to say chaotic. Then the only way to predict the population after a given number of iterations is to actually perform them. This is the "hell" of chaos. There is no shortcut way to predict the future of a chaotic system. Yet it is completely deterministic. If one begins with the same growth rate and start value and does the same number of iterations, the result is always the same. Here we have $r = 2.8$ and start at $x = 0.1$:

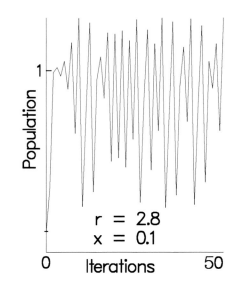

But make a slight change. Here we start again with $r = 2.8$ but $x = 0.1000001$, a one part per million change:

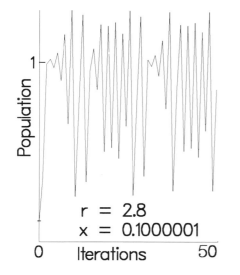

r = 2.8
x = 0.1000001

Notice how this and the previous plot rapidly become different even though r is the same. In the periodic regime of r we could start anywhere, and end up in the same place on the same cycle. In the chaotic regime this slight difference in initial condition gives a drastically different result. *Sensitivity to initial conditons* is a basic property of chaos.

On the next page is a plot of the values that x is attracted to versus r. Since nothing interesting happens below $r = 2$, the range is 1.9 to 3. It is constructed by an attraction process. For each value of r, several hundred iterations of the equation are done to get through the transient behavior of starting.

Then several hundred more iterations are done and the value of x is plotted each time. In the stable cycle region, the iterations all fall on top of 1, 2, 4, 8... discrete points, while in the chaotic region they don't, and we get the "pepper and salt" effect, another fractal dust. This plot is called a *bifurcation diagram*. In the chaotic region there are windows where the chaos disappears. Inset is an enlargement of a piece of one of the windows—again self-similarity!

This diagram turns out to apply not just to the dynamics of populations, but to be rather universal. The period doubling bifurcations are found in the ripples and waves of transitions to turbulent flow, nonlinear oscillators, and a whole gamut of other chaotic systems. This is remarkable in that the physics of turbulence is hydrodynamics, and what does that have to do with dynamics of population growth? The word has already been appropriated, but one might like to call chaos theory a physics of physics, a *metaphysics*.

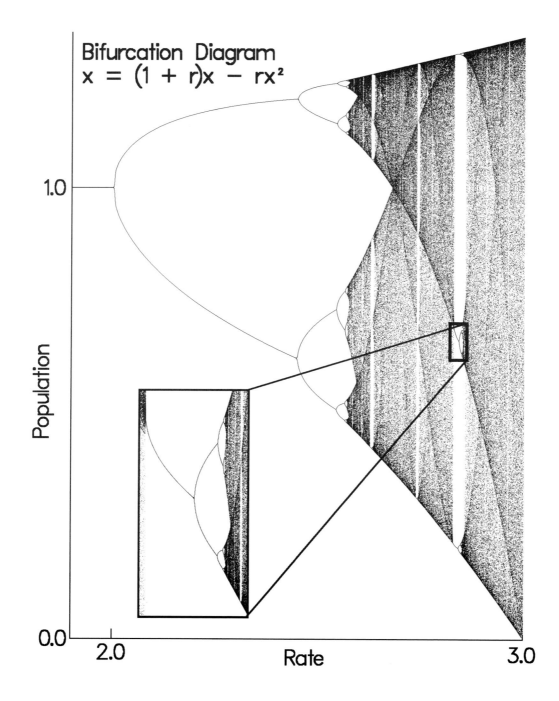

Bifurcation Diagram
$$x = (1 + r)x - rx^2$$

Population

1.0

0.0

2.0   Rate   3.0

96

The diagram is a playground for numerology.

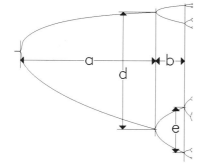

Where *r* is very close to the chaotic region the limiting value of the ratio of successive distances between bifurcations as in *a* to *b* above becomes a universal constant number 4.6692... as does the ratio of successive amplitudes at bifurcation as in *d* to *e* which is 2.5029... These numbers were discovered by Mitchell Feigenbaum, who most importantly realized their universality, in that they are independent of a particular physical system.

If we apply a transformation to the logistic equation—geometrically an affine transformation consisting of a mirror inversion, move, and scaling—we get

$$x_{new} = x^2_{old} + c,$$

which is just the real part of the Mandelbrot equation. There is a bifurcation diagram for it where c is equivalent to r. It has similar period bifurcations and self similar struc-ture. If we superpose it on a similarly scaled Mandelbrot set, the correspondences between its bifurcations of cycles and the regions of the Mandelbrot set along the real axis are obvious. This includes the window in the chaos and the miniature copy of the set out on its "antenna."

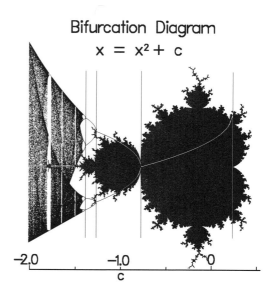

Bifurcation Diagram
$$x = x^2 + c$$

There is much more to chaos than what we have seen above, but there are common threads—complexity that arises from iterating simple rules, sensitivity to inital conditions, and behavior that is in some sense fractal—*strange attractors*. These attractors are not just single points or cyclic arrays of points, but are fractals in the space of the behavioral variables of a chaotic system. The strangeness is their fractal properties. The name predates the popularity of the word fractal. Shown below is a Lorenz attractor at two different stages of iteration. It threads around and back and forth in this three-dimensional folded figure-eight butterly, never intersecting itself, packing its infinite fractal structure into this finite space.

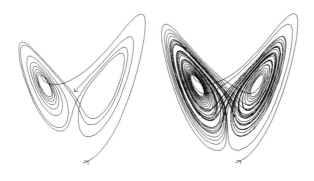

It comes from a computational experiment in long-range weather forecasting. The variables of the space relate to convection in the atmosphere. Three simple equations describe the process. For any given starting point their solution converges onto and remains on the attractor as they are iteratively solved. But a very small differernce in the starting point will give after a few iterations a very different position on the attractor. This is very discouraging for numerical weather forecasting because of the inevitable error and finite precision of the atmospheric measurements that would specify the starting point. Edward Lorenz, the discoverer, calls this the "buttterfly effect" because in principle the flap of a butterfly's

wing could initiate a hurricane.

This is a dramatic way of describing sensitivity to initial conditions. But just which one was the Hurricane Hugo butterfly is another question. Chaos has two faces. On the microscopic scale there is this butterfly effect instability, while on the macroscopic scale there is a great stability, these strange attractors. Just where a chaotic dynamical system will end up on its attractor is exquisitely sensitive to intial conditions, but that it will be pulled onto its attractor, regardless of initial conditions, is certain.

Since the subject is weather, it is natural to wonder what this might mean for the global climate. On the one hand small increases in greenhouse gases may trigger a chaotic climatic catastrophe. On the other, the climate seems to be some sort of an attractor. If that attractor all lies within a region of benign values of the climatic variables, then almost nothing we do may cause much trouble. But it is not clear that there is any less risky way of finding out whether it does than this global experiment we are doing.

Cascade, Tunnel Creek, Big Sur, California 99

Bristlecone Pine, Wheeler Peak, Nevada

Caples Creek near Carson Pass, California   **101**

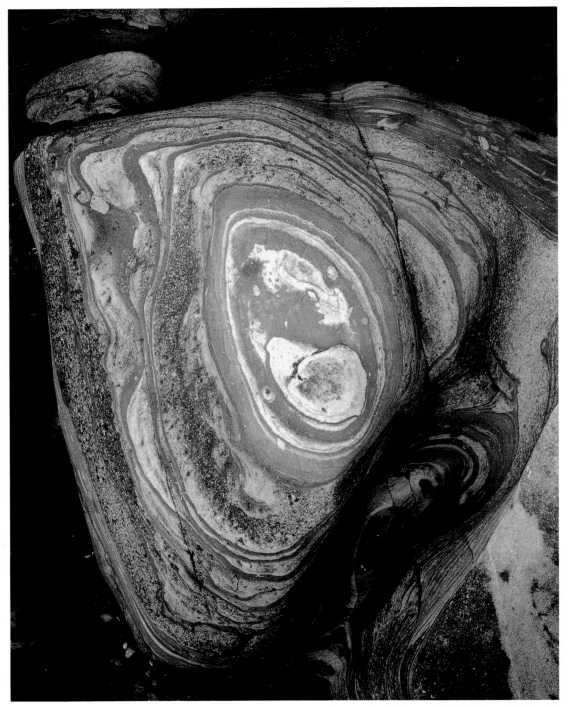

Rock Form #3, Pt. Lobos, California

Antelope Canyon #2, Arizona   **103**

Grass, Pt. Reyes, California

Anise, Big Sur, California

Corn Lily, Cascade Pass, Washington

Prickly Pear, Carmel Mission, California   107

Pahoehoe #4, Hawaii

Fractal geometry as an idea with a name is quite new. While its roots go back a century or more, Benoit Mandelbrot made the essential connection between a variety of disparate mathematical ideas and the irregular, broken, yet patterned aspects of the real physical world. Democratization of computer graphics has had much to do with its popularity and success. Interesting and intriguing work of research quality has been done with personal computers. Yet artists have always been keen observers of the natural world. And fractals are there in the works of artists of all ages. Since the act of drawing or painting is a right-brain process, perhaps even anti-verbal, there was no need for a name. But at times and places there has been a particular consciousness of them, and in some sense a naming of them. E. H. Gombrich in his *Art and Illusion* cites the 1785 work of Alexander Cozens, who advocated "blotting"—the use of accidental ink-blots for the suggestion of landscape. This in turn, Gombrich points out, derives from Leonardo, who wrote:

"You should look at certain walls stained with damp, or at stones of uneven color. If you have to invent some backgrounds you will be able to see in these the likeness of divine landscapes, adorned with mountains, ruins, rocks, woods, great plains, hills, and valleys in great variety..."

The fractal character of Leonardo's sketches of turbulent water is obvious. A body of art with strong and consistent fractal content is classical Chinese and Japanese landscape painting. Typical vignettes from such painting, shown below and on the next page, reveal a particularly clear sense of scaling self-similarity.

Early Western criticism of this body of art was struck by this. It recognized the formulaic or algorithmic aspect but affected to find it sterile, missing its connection to the natural world.

Henry P. Bowie, the author of *On the Laws of Japanese Painting,* is quite direct in describing scaling algorithms for painting various aspects of landscape. A couple of examples that produce a fractal effect for rocks and ledges are "the peeled hemp bark method"

and "the large and small axe strokes on a tree."

Here, scaled aggregations of half-circles become a tree branch.

As he says,

"Nothing is more interesting than such art resources whereby the sentiment of landscape is reproduced by thus suggesting or symbolizing many of its essential features."

Fractal geometry is essential in that sense. And to the comment that it all seems rather mechanical, he replies that for a master,

"It is such second nature for him to employ them that he does so as unconsciously as one in writing will invoke the rules of grammar."

Many of the works of M. C. Escher have a very conscious fractal content. Obvious examples are his circle limit graphics, which imply self-similarity down to the infinitesimal scale. More interesting from my point of view is the consciousness of fractals that appears in his treatment of natural details in his more conventional landscapes. For example see, "Castrovalva" or "Dew Drop" in *The Graphic Work of M. C. Escher*.

As the examples have shown, fractal geometry comprehends the irregular yet patterned aspects of nature in a way that conventional geometry could never do. And it seems to be bound up in the way that visual perception works. In 150 years of photograpy of nature and landscape, there are many examples of high sensitivity to this geometry of nature. Ansel Adams wrote about photographing textures in a way that indicated a deep and implicit understanding of the fractal concept, lacking perhaps only the word we now use. This was in his Basic Photo Series book, *Natural Light Photography,* published in 1952.

"Consider photographing rocky landscapes, another instance that presents serious problems in textural rendition. Beyond a certain distance, a great field of granite boulders will appear as perfectly smooth stones, the natural textures being beyond the resolving power of the lens and/or the emulsion. In order to suggest the substance of these stones it is necessary to include in the very near foreground a boulder in which the texture is adequately revealed. You can then say that the photograph 'reads well.' While you cannot see the texture in the distant boulders, you can see it in the near boulder, and you assume that all the boulders are the same material. It is this awareness of substance that is of vital importance in interpretative photography. The photograph no matter what its function must 'read' clearly."

Mount Williamson, Sierra Nevada, from Manzanar, California
Photograph by Ansel Adams, 1945. Copyright (c) 1991 by the
Trustees of the Ansel Adams Publishing Rights Trust.
All rights reserved.

Cascade near Colby Meadow, Sierra Nevada, California

Cypress Grove, Pt. Lobos, California   115

Desert Holly, Death Valley, California

Buck Creek, North Cascades, Washington

Below Cathedral Rocks, Yosemite, California

Lodgepole Pines, Sierra Nevada, California 119

Bamboo Forest, Road to Hana, Maui

Merced River, Yosemite, California

Sage and Blackberries, San Mateo, California

The fractals we see in the natural world and the chaotic behavior we observe combine complexity and simplicity in a complementary way. There is, of course, much complexity in the world. The surprise is how much of it can arise, not from other complexity, but from the iteration of simple rules, simple processes. Traditionally we speak of the polar opposites, order and chaos (in the old complete randomness sense). The new kind of chaos, this thing that is in some way fractal, is not just something in the middle, a splitting of the difference; it is a third pole.

There are many analogies. An underlying one, with much explanatory power for the rest, is to cast things in terms of information theory and *complexity,* whose definition we will sharpen. Information in this sense is an accounting of the number of possibilities, states, or conditions that a system could be in. Total order has negligible information content; there is only one state available. Total chaos (old sense) is a maximum of possibilities, thus maximum information content. This third pole lies between. What distinguishes it? The answer is complexity in the sense of the amount of computation, the amount of effort it takes to describe what is there. Total order is obviously null complexity. Total randomness is also really null complexity. The average of a large number of fair coin tosses scored 1 for heads and 0 for tails is 0.5000000.... But where there is deterministic chaos, complexity is at maximum. There is no computational shortcut: every iteration must be done.

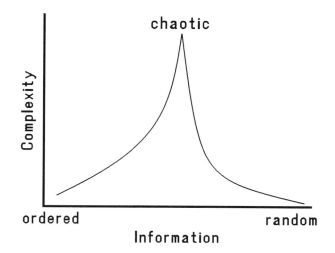

A second analogy is found in the physics of certain phase transitions. A phase transition is a change in the state of matter, such as from solid to liquid or liquid to gas. A magnetic material undergoes a phase transition and becomes non-magnetic when heated above its *Curie* temperature (named after Pierre Curie, husband of Marie Curie). Absolute zero temperature corresponds to total order, while at infinite temperature there is complete randomness. But at the Curie temperature a maximum in complexity is required to describe the magnetic structure—again the filigreed entanglement, the diversity that flourishes at boundaries.

Claude Shannon, the inventor of information theory, cautioned against extending it to fields where it could not be established on a rigorous basis. But the extension of metaphor and analogy yields insight. With that permission, there are triads that make sense. Between a monotone and white noise lies music, and the best of it is at this peak of complexity. Between totalitarian order and anarchy lie reasonable social and political systems, and the best of them encompass a peak of ordered diversity. It is likewise with visual images. My personal approach to photography is to seek out and explicitly display complexity as best I can. Other makers of images are more subtle in their approach, yet more securely astride this peak of complexity than I am; I notice a certain bias in my failures toward the pole of randomness. The mathematics of the magnetic phase transitions regards zero and infinite temperature as attractors, while the Curie point is the opposite, a *repeller*. It is hard to stay at that point.

In making photographs of nature, I use my sense that simplicity and complexity are complementary. Complementary though they may be, there is tension between them that can enliven images. It is as if I wouldn't attempt to grasp that complexity without having a conviction of the underlying simplicity.

Mathematics and geometry have influenced photography before. In particular there was a book, *The Curves of Life* by Theodore Andrea Cook, published in 1914. Its emphasis was on spiral growth patterns and the related Fibonacci series. Edward Steichen's photographs of sunflowers and Edward Weston's nautilus shells are well-known examples of its influence.

A result that I hope for from this essay may be like what Eadweard Muybridge did over a century ago. With high-speed flash, he showed with a photograph that at a point in its stride, all four feet of a galloping horse were off the ground. This settled a bet, and from then on painters of galloping horses followed suit. Go out and see fractals!

124

Stranded Kelp, Pt. Lobos, California

Washingtonia Palm #1, Anza Borrego, California

Buckskin Gulch, Arizona

Ocotillo, Anza Borrego Desert, California

Agaves, Anza Borrego Desert, California

Aspens, Hope Valley, California

130

Aspens, Lundy Canyon, California

Oak #2, Arastradero Preserve, Palo Alto, California

Bristlecone Pine, White Mountains, California

Aspens, Boulder Mountain, Utah

134

Hillside, Big Sur, California

135

Oak Grove, Coe Ranch, California

Pine Forest, Pt. Lobos, California

Heracleum, Pt. Reyes, California

138

Fractal Cypress, Pt. Lobos, California

Buckeye Tree, Pt. Reyes, California

Oak #3, Arastradero Preserve, Palo Alto, California

Sand Crab Excavations, Queensland, Australia

Oak Tree, Coe Ranch, California **143**

Ferns, Pt. Reyes, California

Sea Palms and Shell, Pt. Lobos, California

Oak #4, Arastradero Preserve, Palo Alto, California

Stranded Kelp #2, Pt. Lobos, California

Thunderstorm, Grand Canyon, Arizona

Rock Forms #4, Pt. Lobos, California

Waddel Creek, Big Basin, California

Waterwheel Falls, Yosemite, California

Sycamores, Livermore, California

Bridalveil Creek, Yosemite, California

Pandanus, Kauai

Rain Forest, Road to Hana, Maui **155**

Pandanus Roots, Maui

Washingtonia Palm #2, Anza Borrego, California 157

Pandanus, Keanae, Maui

Strangler Fig #2, Queensland, Australia

Bibliography

This is an annotated bibliography of books and articles on fractals and related matters which were useful, helpful, or inspiring in writing this book. Much of it is at a rather technical level, but some of that is visually quite interesting. They are not necessarily the original source, but what was available and comprehensible to me.

Books:

The Fractal Geometry of Nature, Benoit B. Mandelbrot, W. H. Freeman and Co., San Francisco, 1982. Also an earlier version, *Fractals, Form, Chance and Dimension*, W. H. Freeman and Co., San Francisco, 1977. This is the source of everything fractal. It is a widely discursive essay. A considerable amount is quite accessible, but some parts are heavy going.

The Beauty of Fractals, Images of Complex Dynamical Systems, H. O. Peitgen and P. H. Richter, Springer-Verlag, Berlin and Heidelberg, 1986. It contains incredible color pictures of the Mandelbrot set and other iterated systems, high-level discussions of the mathematics of these images, and some philosophical discussions of their meanings.

The Science of Fractal Images, H. O. Peitgen and D. Saupe, eds., major articles by M. F. Barnsley, R. L. Devaney, B. B. Mandelbrot, H. O. Peitgen, D. Saupe, and R. F. Voss, contributions by Y. Fisher and M. McGuire, Springer-Verlag, New York, 1988. The emphasis of this book is the computer graphics and programming of fractals. This present book arose from my essay, "An Eye for Fractals," in it. The graphics in this book owe much to other articles in it.

Chaos, Fractals and Dynamics, Computer Experiments in Mathematics, Robert L. Devaney, Addison-Wesley, Menlo Park, Calif., 1990. This little book shows that much can be taught and understood of the mathematics of fractals and chaos at the level of high school students. It includes simple programs in BASIC for generating fractal graphics. My discussions derive from it in a number of places. This is a good first place to go for more on the mathematics of fractals.

An Introduction to Chaotic Dynamical Systems, 2nd ed., Robert Devaney, Addison-Wesley, Redwood City, CA, 1989. This is Devaney's big book on the subject—suggested for mathematically mature audiences.

Metamagical Themas, Douglas R. Hofstadter, Basic Books, New York, 1985. Chapter 16 on "Mathematical Chaos and Strange Attractors" goes fairly deep into the bifurcation road to chaos while remaining comprehensible to the lay person. Chapter 13 on "Pattern, Poetry, and Power in the Music of Frédéric Chopin" is an appreciation of this music in the spirit of the word *complexity,* as I use it in this book.

Chaos, Making a New Science, James Gleick, Viking Penguin, New York, 1987. This is a very accessible, readable account of the research and the personalities involved in the discovery of deterministic chaos and its fractal geometry.

Fractals, Jens Feder, Plenum, New York and London, 1988. This is an introduction to fractals by a physicist. While its level is quite technical, it has a different flavor from the mathematicians' presentations. It covers and collects in one place some useful background otherwise dispersed in many journal articles, some rather hard to obtain.

Fractals Everywhere, Michael Barnsley, Academic Press, San Diego, 1988. This is the basis of the transformation approach to fractals. The set theory background is elaborated in technical detail.

Lindenmayer Systems, Fractals and Plants, Przemyslaw Prusinkiewicz and James Hanan, Springer-Verlag, New York, 1989. This is an approach to fractals and plant growth based on language theory. The fractal bush in the text is from an algorithm and parameters suggested in this book.

Art and Illusion, A Study in the Psychology of Pictorial Representation, E. H. Gombrich, Princeton University Press, Princeton, 1969. A reader knowing about fractals as a geometry of nature will find certain parts of it especially resonant.

Growth Morphogenesis, Y. Kawaguchi, JICC Publishing, Tokyo, 1985. This is a book of fantastic fractal creatures with a flavor of marine biology swimming in Kawaguchi's video display.

The Curves of Life, Theodore Andrea Cook, Constable and Co., London, 1914, republished by Dover, New York, 1979. This is an extensive discussion of spiral formations in nature and their influence on art. It is known to have influenced Edward Weston and Edward Steichen.

EW:100 Centennial Essays in Honor of Edward Weston, Peter C. Bunnel and David Featherstone, eds., The Friends of Photography, Carmel Calif., 1986. The essay "Curves of Art" by Mike Weaver discusses the influence of *The Curves of Life* on Weston and Steichen.

On the Laws of Japanese Painting, Henry P. Bowie, P. Elder and Co., San Francisco, 1911, republished by Dover, New York, 1952.

The Graphic Work of M. C. Escher, Grammercy Publishing, New York, 1984.

Drawing on the Right Side of The Brain, Betty Edwards, J. P. Tarcher, Los Angeles, 1979. This book describes the right/left brain model of artistic perception. If one does the drawing exercises, it becomes personal experience. The model has its critics, but they seem irrelevant after one has experienced it.

Articles:

"A hexagonal orthogonal-oriented pyramid as model of image representation in visual cortex," A. B. Watson and A. J. Ahumada, *IEEE Transactions on Biomedical Engineering,* 36, 97—106 (1989). This is a concept of how visual information might be organized in an efficient fractal way in the visual cortex.

"Relations between the statistics of natural images and the response properties of cortical cells," D. J. Field, *Journal of the Optical Society of America,* A4, 2379—2394 (1987). Natural images have the statistical properties of random fractals. The response of the visual cortex seems to match this. It is rather complementary to the previous paper.

"Fractal-based description of natural scenes," A. P. Pentland, *IEEE Transactions on Pattern Analysis and Machine Intelligence,* 6, 661—674, (1984). This is mostly about random fractals as a model of landscape. It contains a proof that two-dimensional images of three-dimensional fractals are also fractal—critical to the thesis of this book!

"Botanical tree image generation," M. Aono and T. L. Kunii, *IEEE Computer Graphics and Applications*, 4, 10-34, 1984. Three-dimensional trees generated by simple recursive rules. The algorithm for the fractal tree in the text came from this.

"Packing it in," I. Peterson, *Science News*, 131, 283—285 (1987). This is a readable account of Barnsley's transformation approach to fractals and its application in image compression.

"Solution of an inverse problem for fractals and other sets," M. F. Barnsley, V. Ervin, D. Hardin, and J. Lancaster, *Proceedings National Academy of Sciences, U. S. A.*, 83, 1975—1977 (1986). This is a mathematical paper which succinctly describes the transformation approach.

"A better way to compress images," M. F. Barnsley and A. D. Sloan, *Byte*, (1/88), 215—223. This is the transformation approach described in elementary programming terms.

"Computer recreations" (column) A. K. Dewdney, *Scientific American* (8/85), 16-24. This article, more than anything else, brought the Mandelbrot set and related matters into the popular computer culture. It could be thought of as a particularly insidious computer virus in that it not only takes over computers (very computationally intensive) but also the minds of programmers with its fascination.

Amygdala, a newsletter on fractals emphasizing the Mandelbrot set and its relatives. Published by Rollo Silver, Box 219, San Cristobal, New Mexico 87564. Mandelbrot means "almond bread" in German, and Amygdala means "almond" in Latin.

"Computation at the onset of chaos," J. P. Crutchfield and K. Young, in *Complexity, Entropy and the Physics of Information, SFI Studies in the Sciences of Complexity*, vol. 8, W. H. Zurek, ed., Addison-Wesley, Redwood City, Calif., 1990. This is the source of the concept of complexity as I have used and interpreted it.

Colophon

My technical approach to photography emphasizes the tools and techniques of the western landscape style. I used some medium format (6×6 cm.—Hasselblad), but mostly large format (4×5 in.—Linhof), for the pictures in this book. Fractals have detail on all scales. Large format most of all favors preserving this detail. Ansel Adams' zone system is for some a religion, but for me it is an understanding that lens, camera, film, developer, enlarger, paper, etc. together make up a system for making images, and are not to be considered in isolation from each other.

Most of the graphics in this book were done with a Hewlett Packard Vectra PC using various versions of Borland's Turbo C programming environment. For printing and proofing, I used a Hewlett Packard LaserJet in conjunction with a Tall Tree Systems JLaser expanded memory board. With a direct video connection to the laser printer, it enables instant printing of high-resolution bitmaps created in expanded memory. I assembled the book using Aldus Pagemaker (PC version). The typeface is ITC Garamond from Bitstream. I hope soon to publish the collage theorem transformation editor software.